BABY ANIMALS

BABY FLAMINGOS

by Julie Murray

Cody Koala
An Imprint of Pop!
popbooksonline.com

Hello! My name is
Cody Koala

This book is filled with videos, puzzles, games, and more! Scan the QR codes* while you read, or visit the website below to make this book pop.

popbooksonline.com/baby-flamingos

*Scanning QR codes requires a web-enabled smart device with a QR code reader app and a camera.

abdobooks.com
Published by Pop!, a division of ABDO, PO Box 398166, Minneapolis, Minnesota 55439.

Printed in the United States of America, North Mankato, Minnesota.
102023
012024
THIS BOOK CONTAINS RECYCLED MATERIALS

Cover Photo: Shutterstock Images
Interior Photos: Shutterstock Images; Getty Images
Editors: Elizabeth Andrews and Grace Hansen
Series Designer: Candice Keimig

Library of Congress Control Number: 2023938790

Publisher's Cataloging-in-Publication Data
Names: Murray, Julie, author.
Title: Baby flamingos / by Julie Murray
Description: Minneapolis, Minnesota : Pop!, 2024 | Series: Baby animals | Includes online resources and index.
Identifiers: ISBN 9781098245214 (lib. bdg.) | ISBN 9781098245771 (ebook)
Subjects: LCSH: Animal babies--Juvenile literature. | Animals--Infancy--Juvenile literature. | Flamingos--Juvenile literature. | Birds--Juvenile literature. | Birds--Behavior--Juvenile literature.
Classification: DDC 591.39--dc23

Table of Contents

Chapter 1

Fluffy Gray Feathers

Baby flamingos are called chicks or flaminglets. They hatch from eggs. A female flamingo usually lays just one egg at a time.

Watch a video here!

A baby flamingo has fluffy gray feathers. Its legs are pink and **swollen**. It has a reddish **bill** that is straight.

A newly hatched flamingo chick is the size of a tennis ball.

Both parents feed their chick **crop milk**. This food comes from the parents' **digestive tracts**. It is bright red. It helps the chick stay healthy and grow.

Chapter 2

Growing Up

Baby flamingos can walk within their first few days of life. They begin to leave the nest to **socialize**. Their parents keep a close eye on them.

Learn more here!

Young flamingos are gathered in a group. It is called a creche. A few adults watch over the group.

Parents continue to come and feed their baby. They find each other by using their voices.

As baby flamingos grow, they learn how to find their own food. Their feathers begin to change. Their **bill** starts to curve downward. Their legs grow long. They start living on their own at around 12 weeks old.

Flamingos can grow to be five feet (.3m) tall.

Chapter 3

Pink Feathers

It takes one to two years for flamingos to get the full pink color in their feathers. The pink color comes from the food they eat.

Explore links here!

Flamingos mainly eat algae and brine shrimp. These foods are high in a **pigment** called beta-carotene. It gives flamingos their color!

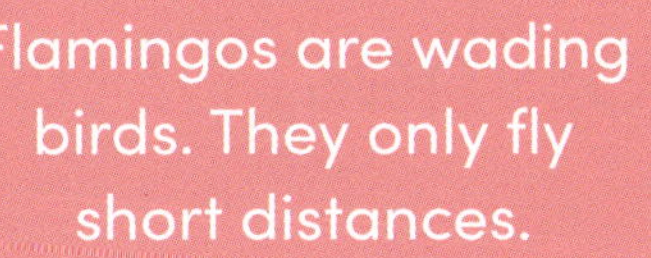

Flamingos are wading birds. They only fly short distances.

Chapter 4

Part of a Colony

Flamingos live in large groups called colonies. Each colony can have a few birds or tens of thousands! They live in wetland areas such as lakes, marshes, or lagoons.

Where Flamingos Live

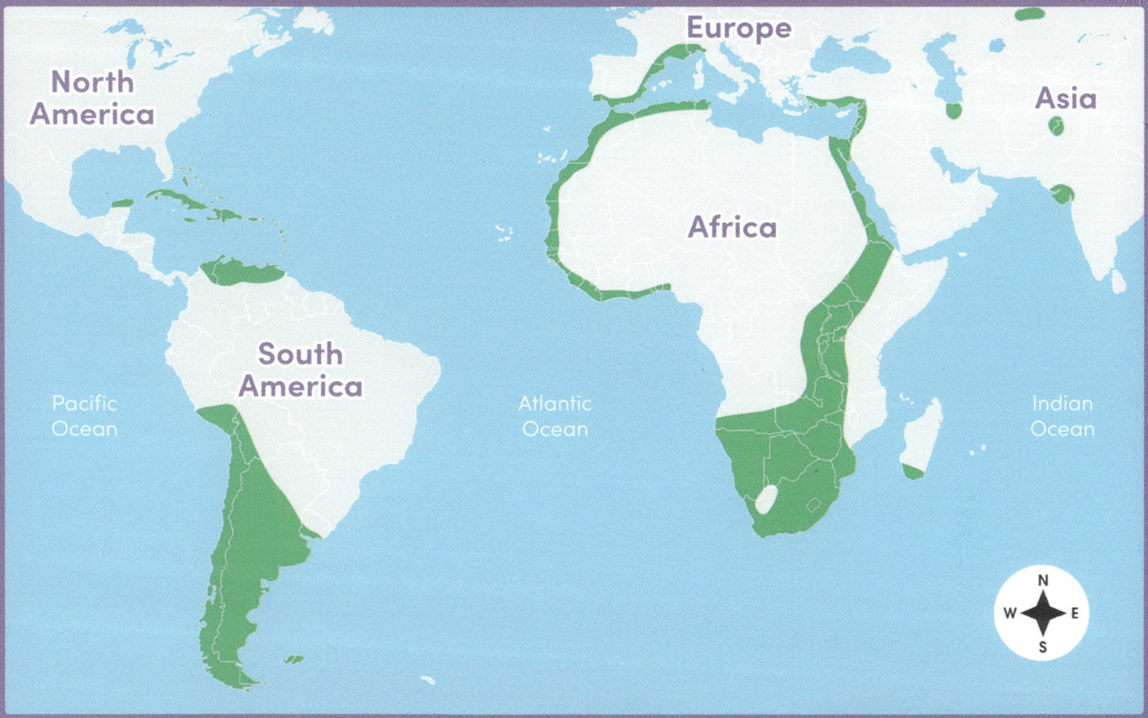

North America

Europe

Asia

Africa

South America

Pacific Ocean

Atlantic Ocean

Indian Ocean

N W E S

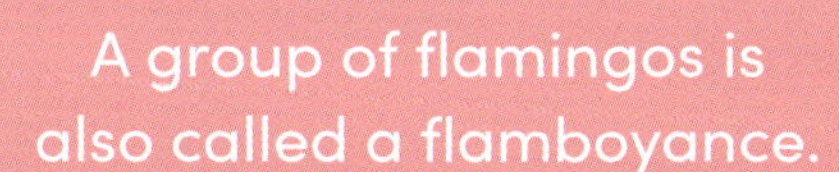

Complete an activity here!

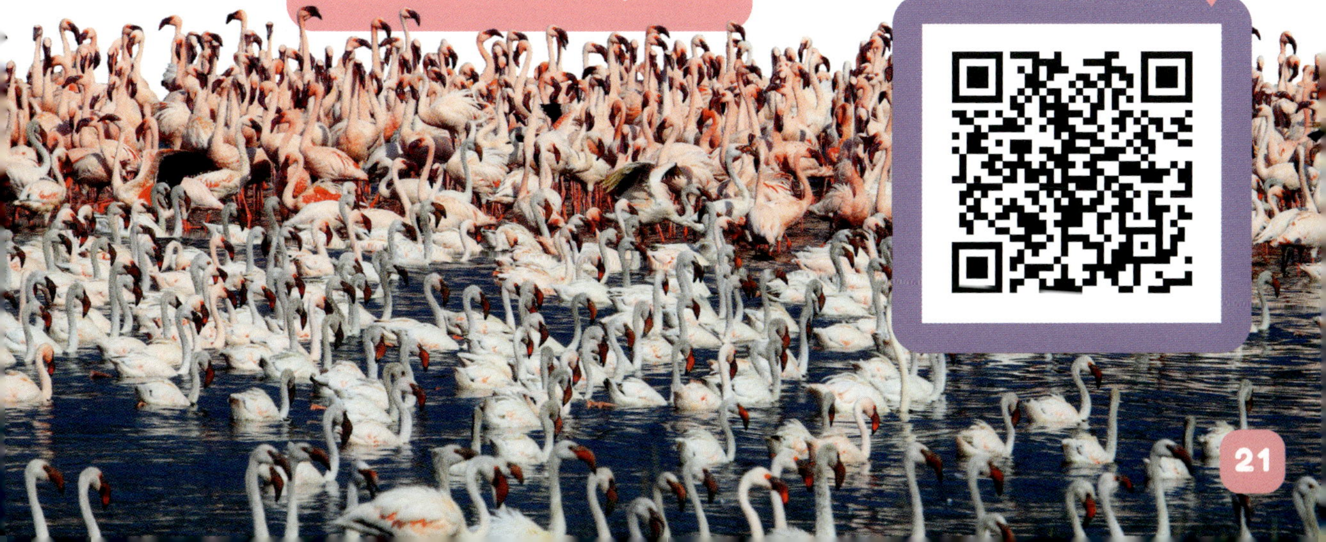

Making Connections

Text-to-Self

What is one new thing you learned about baby flamingos from this book?

Text-to-Text

Have you read books about other birds? How are those birds similar to or different from a flamingo?

Text-to-World

Baby flamingos and adult flamingos look very different. Can you think of another animal that changes a lot as it grows?

Glossary

bill – the parts of a bird's jaw that form the beak.

crop milk – a semi-solid secretion that is made in the crop of certain parent birds to feed their young. A crop is the pouch located on the front of a bird's neck.

digestive tract – the path of hollow organs that move and process food and liquids through the body.

pigment – anything that is used to or serves to provide color.

socialize – to engage in social activities.

swollen – made larger by inflation or bulging.

Index

Online Resources

popbooksonline.com

Thanks for reading this Cody Koala book!

This book is filled with videos, puzzles, games, and more! Scan the QR codes* while you read, or visit the website below to make this book pop.

popbooksonline.com/baby-flamingos

*Scanning QR codes requires a web-enabled smart device with a QR code reader app and a camera.